AKLILU LEMMA

The Story of a Young Scientist and a Magical Plant

To request permission, contact the publisher at info@lella-menged.com.

Illustrated by Yorris Handoko

Hardcover ISBN: 978-1-959202-05-9
Paperback ISBN: 978-1-959202-04-2

Library of Congress Control Number: 2022924014

Lella Menged LLC
5900 Balcones Drive
Suite 100
Austin, TX 78731

www.lella-menged.com

Dedication

For my mother, Aberash Alemu, who is my moral compass and spiritual guide, for teaching me to love my country, its heritage, and its people.

For my father, Aweke Embiale, who is my role model and who instilled in me the importance of having a vision and working hard.

Acknowledgments

I want to thank all of you who have contributed to this children's book in one way or another. I want to thank Biniam Hirut for his support in gathering information about Professor Aklilu Lemma. I also want to thank Fanuel Zewdu, a medical microbiologist, for his contribution in reviewing the technical parts of the manuscript. I am also grateful for the fantastic job Krista Zimonick did in editing the book.

Phytolacca
dodecandra
(Endod)

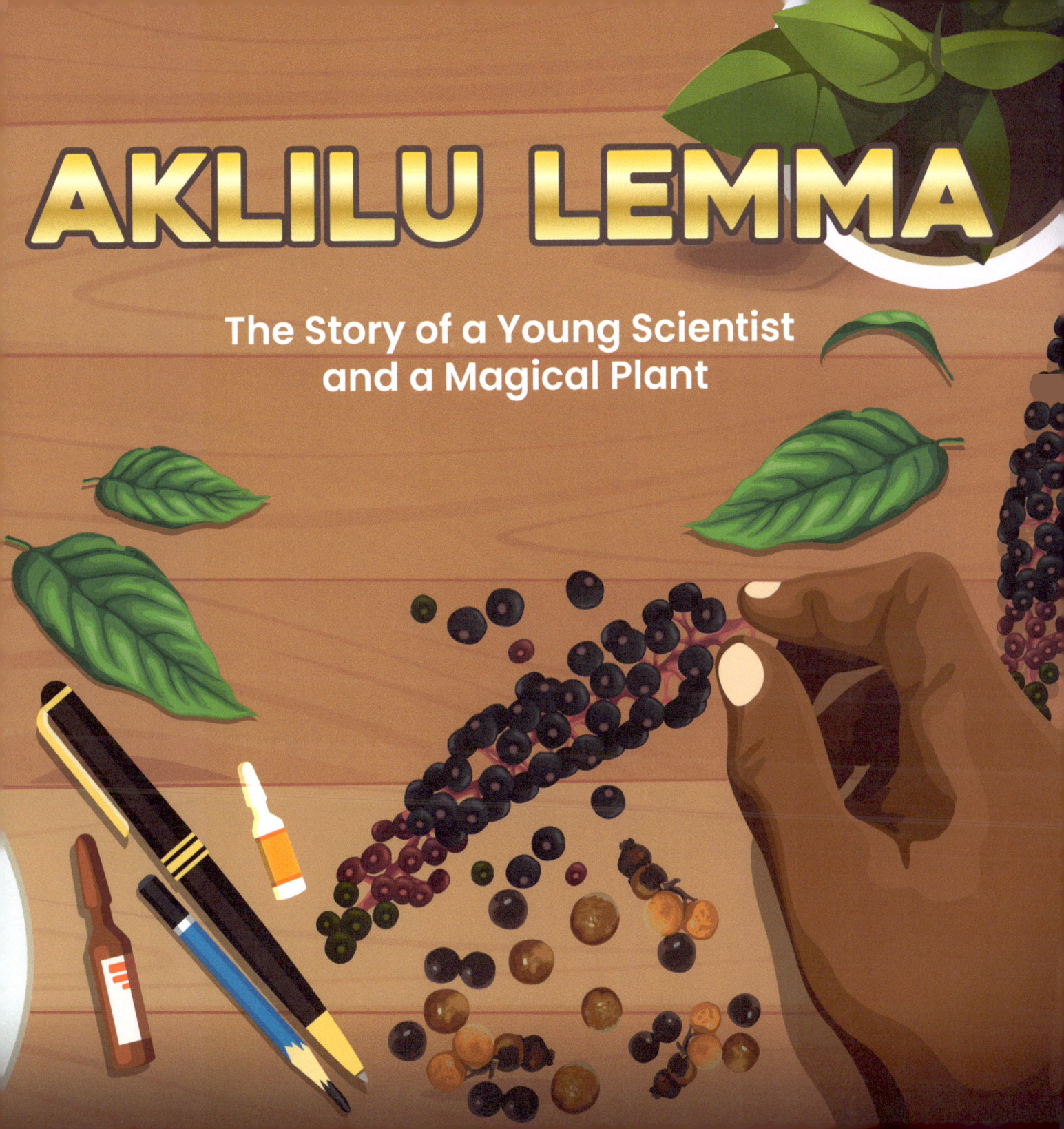

AKLILU LEMMA

The Story of a Young Scientist
and a Magical Plant

History is full of unsung heroes.

There are hundreds, no, *thousands* of people from across the planet and throughout history who have done truly amazing things for the human race…
and most people don't know a thing about them!

This is the story of one of those heroes. A wonderful professor called Aklilu Lemma. He was a *pathobiologist* from Ethiopia, and led an amazing life full of achievements, campaigns, and adventures.

He founded institutes, discovered solutions to big problems, and set up foundations. He was once a science and technology advisor to the Ethiopian government, and he once received a science gold medal from the Emperor of Ethiopia, Haile Selassie!

And believe it or not, that's only the tip of the iceberg!
Aklilu Lemma managed to achieve *loads* more than that!
These aren't even his greatest achievements.

And speaking of greatest achievements...
his greatest achievement is what
we are really here to talk about.

Very early on in his career, Aklilu Lemma discovered a natural cure to a disease known as *schistosomiasis*, which won him an award called the Right Livelihood Award… also known as the "Alternative Nobel Prize". That was a MASSIVE achievement, especially for someone from Ethiopia!

But look, before we get on to this big discovery of his, let's talk about a few other things first…

Now, first of all, you may have noticed some complicated words on the first few pages. Let's explain some of those, shall we?

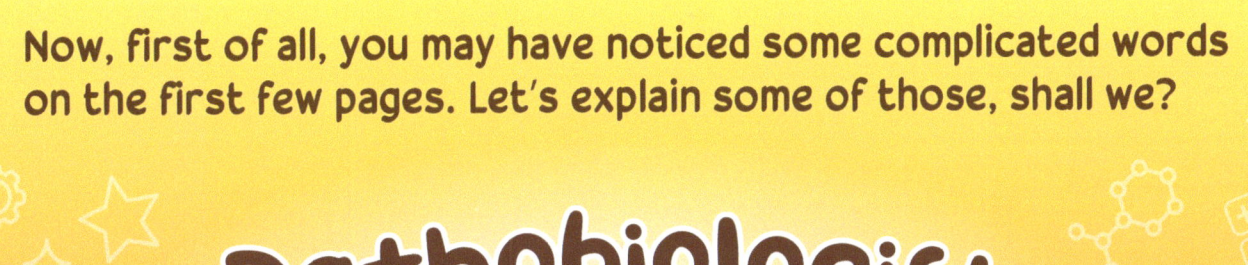

pathobiologist
Schistosomiasis

We explained that Aklilu Lemma was a *pathobiologist*, remember?
'What is a *pathobiologist*?' you might wonder.
Well, a *pathobiologist* is somebody whose job is to research *pathology*.
Pathology is an area of science that looks at all things to do with disease.
What diseases are, how they are created, how they spread
and how they are cured.

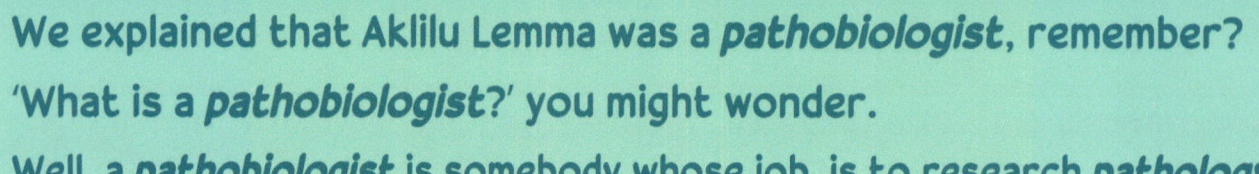

So, it would be fair to say that our friend Aklilu Lemma was an expert on diseases.

Now, we also said that he was most well-known for finding a cure to a disease known as *schistosomiasis*. 'What in the world is *schistosomiasis*?' you might ask. Well, Schistosomiasis, also known as *bilharzia* (another tricky word, I know) is a horrible disease found all across the world.

It was a disease that was killing thousands of people in Aklilu's home country of Ethiopia.

The semi-mature larvae leave snails to infect humans to mature further and produce eggs.

Contaminated fresh water by parasite eggs

Newly-hatched parasite larvae infest snails.

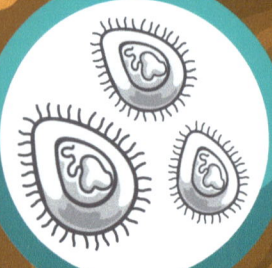

And it was very hard to control it… until he came along, of course!

The disease is caused by a parasite worm (a tiny, tiny creature) called flatworm that lives inside certain types of freshwater snails. The parasites crawl out of the snails and onto human skin. If the parasite gets inside a human, it can make them very sick indeed. Kids that used to swim in contaminated water could be infected. Gross huh?

Enough detail on the disease for now, don't you think?

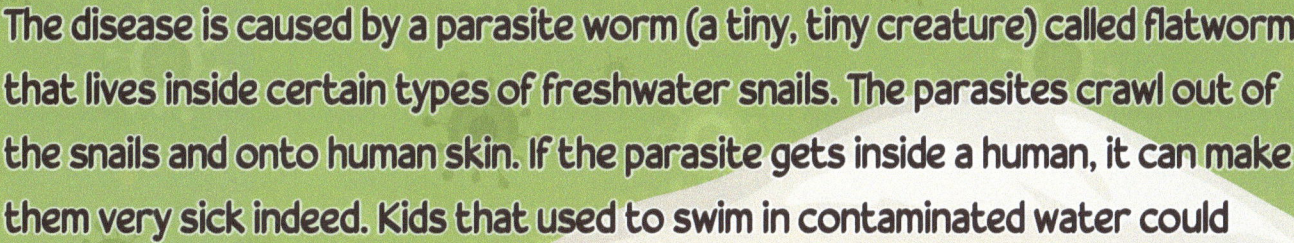

Enlarged liver and spleen

High fever

Diarrhea

Weight loss and loss of appetite

Skin rash

Now that we've explained a few things, let's go back to the beginning. Where did Aklilu Lemma begin his life, and what is the story of his amazing discovery?

Aklilu was born on September 18, 1935, in a city called Jigjiga in the eastern part of Ethiopia. It was a regular, humble town full of ordinary people. Aklilu's parents were named Ayelech Lemma and Ato Bekele Woldeyes. He grew up with his mother's family and took his maternal grandfather's last name.

However, even though he was born in the village, Aklilu was raised in a nearby city, Harar. That was where he finished elementary school. We know that he was a good student who studied hard and always got good grades.

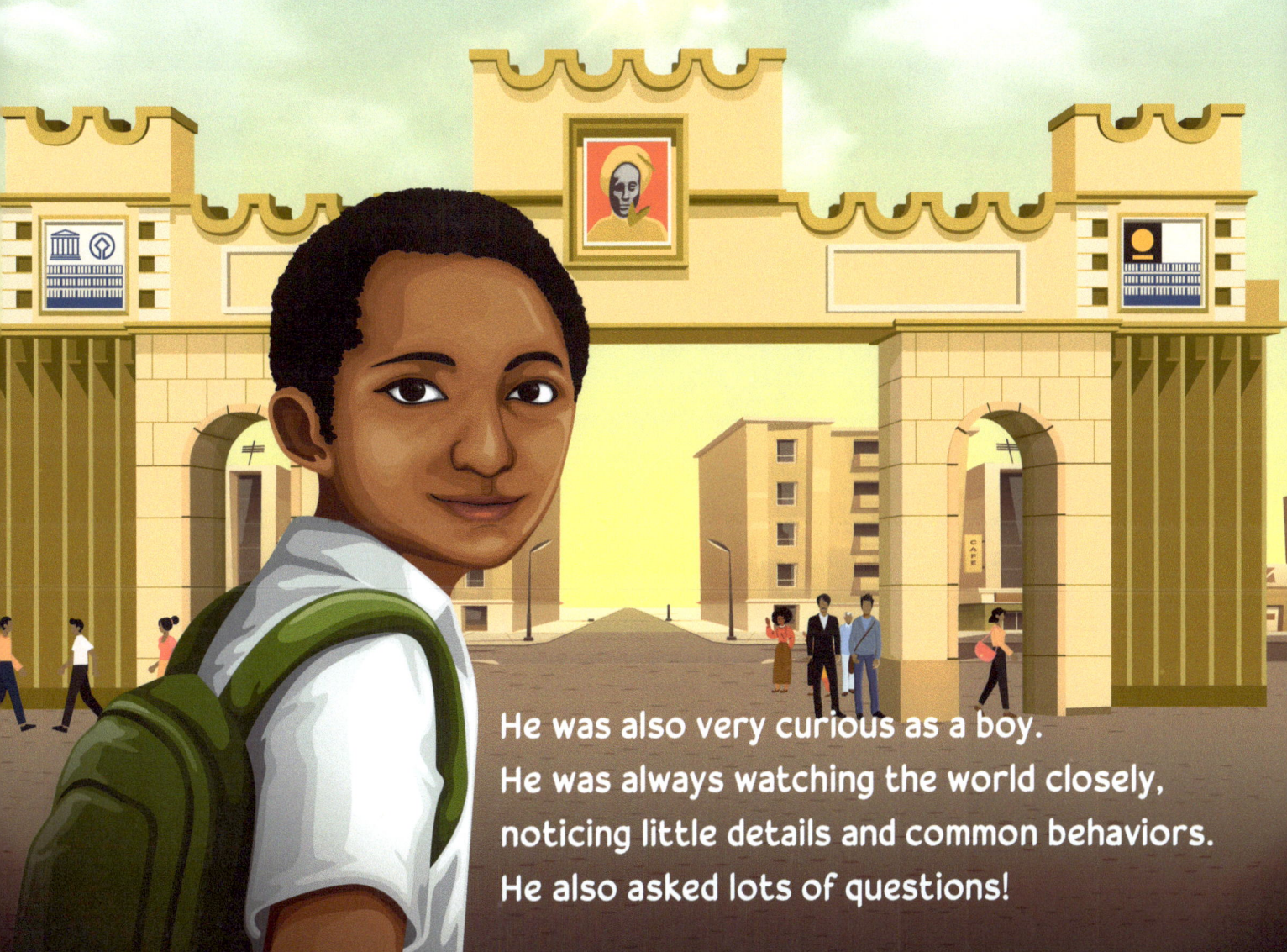

He was also very curious as a boy.
He was always watching the world closely,
noticing little details and common behaviors.
He also asked lots of questions!

After finishing high school, Aklilu then went to university! This was a big achievement for a boy of his background, so he was already doing very well for himself.

He went to Addis Ababa University to study. He studied science and got fantastic grades.

Then he moved to America, to the University of Wisconsin, to earn his master's degree. And *then* he went to a different American university called Johns Hopkins and earned a DSc! (Doctor of Science)

It was when Aklilu was studying at Johns Hopkins that his amazing journey to finding the cure for *schistosomiasis* began.

One day, sometime in the year 1964, he was back at home in Ethiopia. He had traveled to a town called Adwa to investigate the distribution of freshwater snails. He knew that this was important to discovering more about *schistosomiasis*.

He was watching women washing clothes in a local river and noticed something strange when they were done. The part of the river they had been using to do their laundry was full of dead snails . . . Weird, huh?

It was so weird that Aklilu had to work out what the cause was. It turned out that a common plant called *endod* had killed them! How?

Well, *endod* (sometimes called *soapberry*) was what women in the area had been using to clean their clothes for generations. There was something in the plant that was killing the snails. How did Aklilu know this? Well, here's what he said about it all . . .

in areas downstream from where people were washing clothes with the Ethiopian *soapberry* . . . there were more dead snails floating around than anywhere else (including areas where people were washing with commercial soap).

Observing this phenomenon repeatedly, I collected some live snails from upstream and asked one of the women to put a bit of the *endod* suds from her washbasin into the snail container.

Shortly after, the snails shrank, passed a few bubbles of gas . . . and died!.

He knew this plant must contain the answer to *schistosomiasis*!

So, Aklilu got to work. He tested and experimented with *endod* for two whole years! It was in 1966 that he made a major breakthrough.

He found the part of the plant that was responsible for killing snails and flatworms and turned it into a *molluscicide.*

"Excuse me, but what is a *molluscicide*?" you might be thinking. Well, a *molluscicide* is a general word for chemicals that kill mollusks such as slugs and snails.

Today, people use them all the time. They come in sprays, powders, pills, and pellets. However, back in the 1960s, *molluscicides* were unheard of. Aklilu's *molluscicide* was one of the first ever found!

Aklilu presented his discoveries to a science forum in Africa. The scientists there were extremely excited. Why?

Well, first of all, *schistosomiasis* was a BIG problem across Africa. And this discovery was the first real answer to that problem. But not only was Aklilu's *molluscicide* effective, but it was also cheap to make, not toxic, and good for the environment.

But there was still so much to learn about the *endod* plant. Aklilu spent the next two years at Stanford Research Institute in California, looking at the chemicals in the plant and working out the many different ways those chemicals could be used.

Eventually, after loads and loads of experiments and tests, Aklilu managed to work out *exactly* which part of the plant contained the chemical that killed mollusks, and exactly which chemical it was. The chemical was named after him. To this very day, it is known as...

Lemmatoxin

It is found in the fruit of the *endod,* not in the leaves or the stems.

Aklilu and his team also worked out that *Lemmatoxin* helped to control the effect of other creatures that carried disease, not just snails and slugs.

It also worked on mosquitoes, copepods, and black flies.

Snail

Slug

Mosquito

Copepods

Black fly

A trial was done on 3,500 children, and the results were incredible! Before the trials, disease affected fifty in every hundred children. After the trials, disease affected seven in every hundred children.

Awesome huh? Well, yes. But while all this great stuff was going on, there was also some not-so-great stuff going on.

Back in 1964, when Aklilu first made his discovery, loads of people suggested that they help with further research. You see, there were so many different kinds of *endod* from all over the world that for Aklilu to research them all on his own would have been too much.

Besides, Aklilu wasn't one for personal glory. He thought that making discoveries and helping people were much more important.
It didn't really matter to him who made the discoveries, as long as somebody did!

So, he gave a sample of *Lemmatoxin* to the Tropical Plant Products Institute in London. But they were sneaky. They made so many exciting discoveries that they decided to *patent* their research in the UK. This was very upsetting for Aklilu.

Basically, what that meant for Aklilu was that he wasn't allowed to do research on or make things out of the *endod* plant in the UK. Only the Tropical Plant Products Institute could do those things. They completely cut him out!

Imagine that! After all his hard work and research, they didn't want him to be involved or have any credit at all!

So, what did Aklilu do?
Well, he got his own *patent* in the USA. Which meant that he and the Tropical Plant Products Institute were now in competition!

On top of that, the scientists throughout the UK that had been telling the world how amazing Aklilu's discoveries were started saying bad things about him and his research. That he didn't know what he was doing. That his research wasn't good enough. That he didn't do his work properly, so the results were wrong!

This made doing more research really tricky. Aklilu needed funding from the World Health Organization for more research to be done, but they weren't so sure after all the bad stuff that had been said about his research. His funding requests were denied.

But this didn't stop Aklilu, of course. He kept going. As the years passed, he kept studying the *endod*, giving his own time and money.

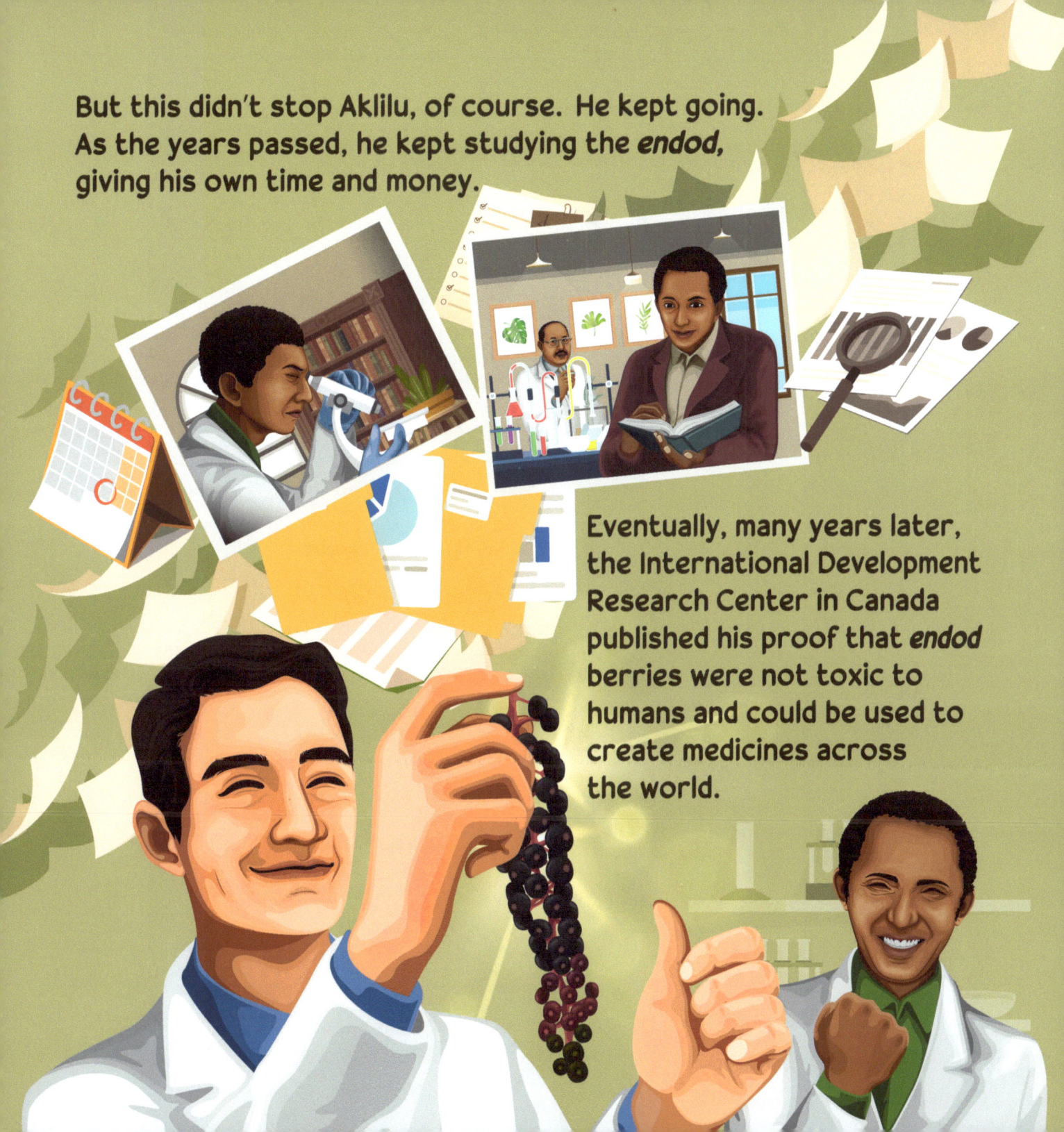

Eventually, many years later, the International Development Research Center in Canada published his proof that *endod* berries were not toxic to humans and could be used to create medicines across the world.

He was eventually rewarded for all of his fantastic work in 1989 when he won the "Alternative Nobel Prize", twenty-five years after his first breakthrough.

What made this an even more exciting achievement was that he won it with his research partner, Dr. Legesse Wolde-Yohannes, whom he had been working with for years. Perhaps we'll talk more about him in another book!

Aklilu remains one of the most successful scientists in all of Ethiopian history.

Fun Facts

Aklilu has three patents, five books, and sixty scientific papers to his name!

The *endod* plant is also used to make soap and shampoo across Africa.

Aklilu had three children.

He died on April 5, 1997, at the age of sixty-two. He was buried in his home country of Ethiopia.

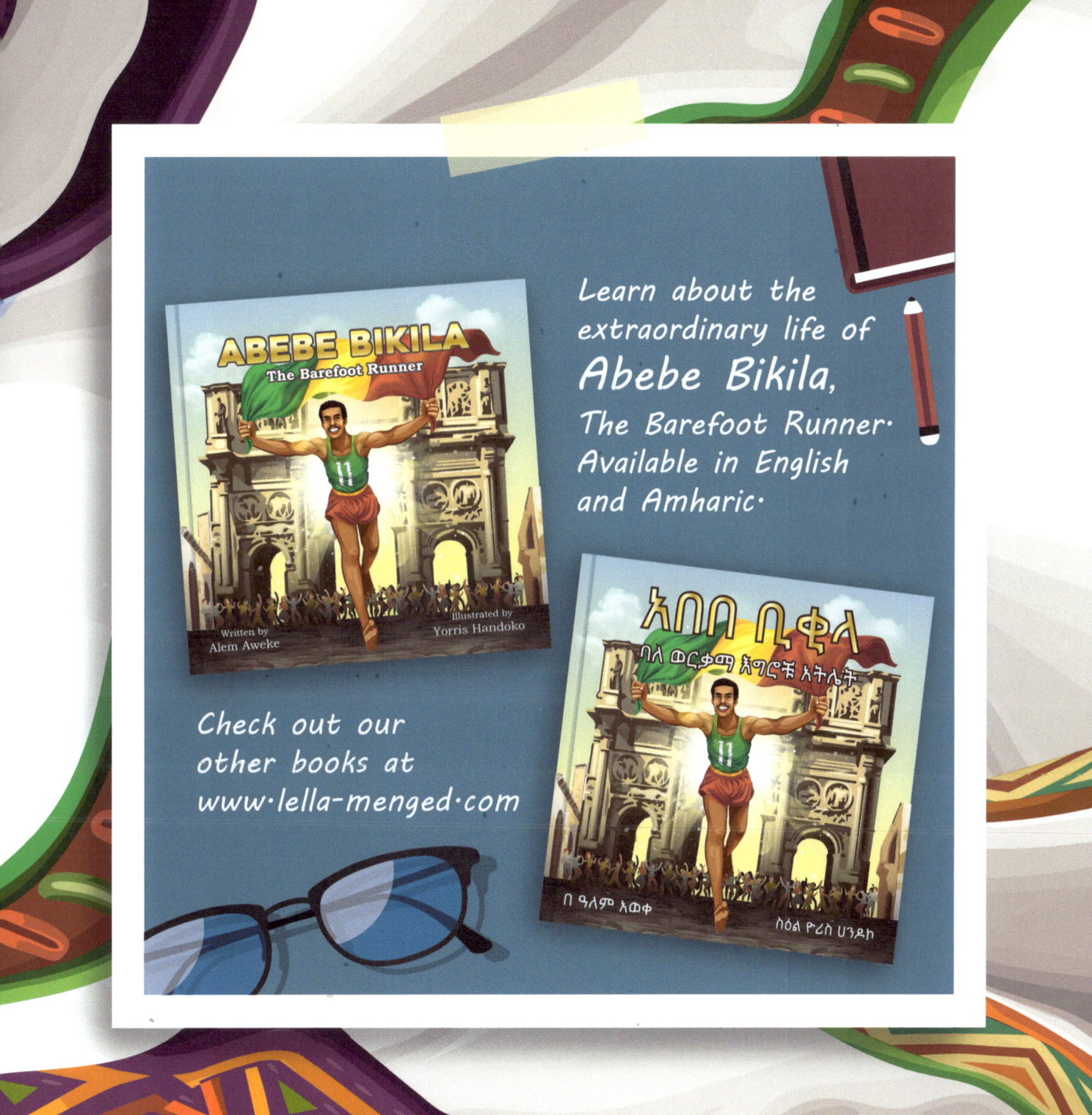

Learn about the extraordinary life of **Abebe Bikila,** The Barefoot Runner. Available in English and Amharic.

Check out our other books at www·lella-menged·com

www.ingramcontent.com/pod-product-compliance
Lightning Source LLC
Chambersburg PA
CBHW041556120626
46551CB00002B/224